ALL NATURAL BEAUTY

ALL NATURAL BEAUTY

Organic & Homemade
Beauty Products

**KARIN BERNDL
& NICI HOFER**

hardie grant books

To Miriam, Sieglinde, Regina, Herbert, Helena & Don

Rainforest, Tobago 2013

CONTENTS

INTRODUCTION

In modern society, we buy our food in supermarkets and our remedies in pharmacies. This dissociation seems entirely normal to us today, but is devastating in truth.

Our plant world has bestowed on us an amazing array of natural foods, herbs and flowers, which can be used both internally and externally in ways that are healing for the body and spirit. Natural ingredients are absorbed easily by our bodies, which can't always be said of synthetically produced substances.

In this book we use herbs, fruits, seeds and natural oils to nourish our skin and bodies with the same care as if we were preparing food for dinner — which is why sometimes you might want to taste the mixtures as you make or use them. We don't blame you, but try and only use the ones on these pages to feed your skin externally!

BEFORE YOU GET STARTED

Make sure you buy essential oil from a trusted source. Use organic ingredients and cold-pressed carrier oils to ensure the best quality for your skin.

Bottles of carrier oils should be kept in the fridge, because as soon as the oil has been in contact with light and oxygen it will start to deteriorate. Refrigeration will slow down this decline in quality.

It is also recommended to store essential oils in the fridge. Be sure to remove oils from the fridge up to 12 hours before use to allow them to adjust to room temperature.

The shelf life of your oils will very much depend on how long the oil has been stored already before purchasing. If the oils are bought from a trusted source, the use-by date will be clearly visible on the bottle. Keep your potions out of direct sunlight, preferably in a cool place.

Grapeseed base oil is a good alternative to nut oils for those suffering from nut allergies.

We use glass instead of metal to heat, mix and stir ingredients, and favour dark glass pots for the storage of our freshly mixed oils and creams.

Use face masks immediately – or store them for a maximum of 24 hours in the fridge.

Be sure to test the ingredients of your homemade potions on small areas of your skin first (the inside of your elbow, for example) to make sure that you are not allergic to any of them. Never apply undiluted essential oils directly to your skin.

The usage and dosage of essential oils differs for children, babies and pregnant women and some oils have restricted use. Please seek expert advice if you are pregnant, using products on children and babies, or have a history of allergies.

These recipes and their oils and essential oils are our personal favourites. You may want to replace some ingredients or swap in oils that are beneficial to you personally. See pages 98-102 for the properties of the ingredients used in this book.

Beeswax melts at 62–64°C (143.5–151°F). It is advisable to use a thermometer to get used to the process. Be careful not to overheat the oils and wax by heating them further than the required melting temperature of the ingredients.

When you mix beeswax with oils, the temperature should not reach above 70°C (158°F). When the beeswax and oils are blended, remove the mix from the heat.

For best results, mixtures that include water (or another liquid such as aloe vera) need to be blended with a stick blender after cooling them down to approximately 50°C (122°F). Add the essential oils only once the mixtures have cooled down to about 40°C (104°F).

Body scrubs can be slippery. Be extra careful not to slip in the bath or shower when using them.

ANTI-AGEING
ROSE
FACIAL OIL

My skin just loves this nourishing facial oil. Its superpower is to fight dry skin and this blend of oils even has anti-ageing properties – what's not to like? So, despite what they say – life's coming up roses!

– NICI

THINGS YOU'LL NEED

2 tablespoons rosehip oil

½ teaspoon avocado oil

6 drops of rose absolute oil

GET STARTED

Simply pour all the oils into a small bottle and give it a good shake until they are well-mixed. And that's it!

Apply this precious face oil every morning and evening to cleansed skin.

If you give the oil to a friend as a gift, it's a nice idea to name it. We suggest 'Magical Superpower Facial Oil', but 'Rose Oil' is an option too.

SUPERPOWERS EXPLAINED

Full of protein and fats, avocado oil is ideal for dry skin. It is rich in vitamins C, E and K and magnesium and potassium. Rosehip oil is very high in important essential fatty acids and helps regenerate damaged skin tissue. It penetrates deep into the skin and stimulates collagen production. It is also moisturising and anti-inflammatory and helps to heal wounds and soften scars. Rose absolute oil is antibacterial, antiviral, antispasmodic, is known to have rejuvenating properties and promotes a glowing complexion.

SHELF LIFE
4 MONTHS
STORE IN A COOL, DARK PLACE

COOLING AFTER-SUN CREAM

Sunburn can be painful and any immediate relief is very welcome. Aloe vera, in combination with these skin-healing additives, will make you feel cooler in no time.

– KARIN

THINGS YOU'LL NEED

1 tablespoon beeswax

50 ml (1¾ fl oz/scant ¼ cup) wheat germ oil

50 ml (1¾ fl oz/scant ¼ cup) jojoba oil

100 ml (3½ fl oz/scant ½ cup) aloe vera gel

1 teaspoon vegetable glycerin

5 drops of lemon juice

18 drops of immortelle essential oil

18 drops of frankincense essential oil

8 drops of peppermint essential oil

6 drops of rose otto essential oil

GET STARTED

Using a double boiler, melt the beeswax gently in a bowl, over a low heat, until liquid. Heat the jojoba oil, wheat germ oil and glycerin in the same way in a separate bowl, stirring gently and being careful not to overheat the mixture. Once the beeswax has reached the same temperature as the oily mixture, mix the two and stir well. Remove from the heat as soon as the ingredients are well-blended. Stir carefully while the mixture cools down. In a separate bowl, heat the aloe vera with the added lemon drops very gently only to about 45 degrees, also using a double boiler. Once the oily mixture starts to become milky, add the warmed aloe vera and blend the solution with a stick blender to form your very own cream. Be sure the aloe vera and the oils have emulsified properly. Once you are happy with the consistency, add the essential oils and stir again. Immediately pour into small pots to keep.

SUPERPOWERS EXPLAINED

Jojoba and wheat germ oils are rich in vitamin E, have anti-inflammatory and wound-healing properties. Aloe vera is anti-inflammatory, antibacterial, antiviral, moisturising and cell-regenerating. Lemon juice acts as a preservative. Immortelle, frankincense and rose otto essential oils help heal wounds and scars and are also cell-regenerating. Peppermint is strongly analgesic, cooling and stimulates the blood flow. Vegetable glycerin is hydrating, beeswax is moisturising.

SHELF LIFE
1 MONTH
KEEP REFRIGERATED

LIPSTICK-READY LIP SCRUB

Dry and cracked lips can be a bit annoying, especially if you want to wear an empowering, kick-ass red lipstick. How I love that strong and confident look. For me, a bright red lipstick never fails to signal, 'Listen up, I have something to say.' But lips that are dry as a snake mid-peel aren't the ideal base for any shade of red. Luckily, this can be fixed with a quick mix of a few simple ingredients that even the barest of kitchens will have stocked (aka, even mine).

– NICI

THINGS YOU'LL NEED

2 tablespoons brown sugar

1 tablespoon honey

1 tablespoon olive oil

GET STARTED

Whisk all the ingredients together. Apply to your lips and give them a gentle scrub until they feel like new. Rinse the scrub off with plenty of water. Apply our moisturising lip balm (page 91) before moving on to the fun bit – choosing your shade of red and being bold.

SUPERPOWERS EXPLAINED

Sugar works superbly as an exfoliant. Honey has wound-healing, anti-microbial and antibacterial properties. Olive oil is antibacterial and anti-inflammatory, protects and nourishes skin and is high in vitamins, antioxidants and minerals.

SHELF LIFE
USE RIGHT AWAY
ENOUGH FOR ONE APPLICATION

SOOTHING
OAK
FOOT BATH

Not so long ago, Celtic druids made their wands from oak, a symbol of endurance and strength. They didn't just appreciate oak trees for their tool-making, but also performed religious rites in oak groves and listened to the trees for divine messages. Back to the here and now — celestial rumour has it that oak bark is still good for a bit of magic: a soothing oak foot bath will be a welcome relief for your sweaty feet!

– KARIN

THINGS YOU'LL NEED

3 tablespoons oak bark

1 litre (34 fl oz/4 cups) cold water

GET STARTED

Bring the oak bark and water to the boil in a pot. Cover with a lid and simmer for 30 minutes on a very low heat.

Prepare a foot-bath with warm water and add the strained oak infusion to the bath. Enjoy for 20–30 minutes, then dry your feet and apply one of your homemade creams to complete the pampering session.

SUPERPOWERS EXPLAINED

Oak bark is antiperspirant, antibacterial, antiviral, anti-inflammatory, astringent and alleviates itching.

SHELF LIFE
1 DAY
ENOUGH FOR ONE SESSION

CHOCOLATE MOUSSE BODY CREAM

THINGS YOU'LL NEED

4 tablespoons cocoa butter

2 tablespoons shea butter

2 tablespoons coconut oil

2 teaspoons cocoa powder

1 teaspoon ground cinnamon

20 drops of eucalyptus essential oil

My fond love for dessert doesn't stop with sweet indulgences after every meal. Oh no, it goes much further: in the form of this delicious-smelling chocolate mousse-like treat, which is not meant to be eaten - even though it looks so tempting!

- NICI

GET STARTED

First things first, melt the cocoa butter and shea butter in a double boiler. Stir them a little and once they are nearly melted, add the coconut oil. Once all three are completely liquid, take off the heat and mix in the cocoa powder and cinnamon.

Finish with the addition of the eucalyptus essential oil. Really concentrate so you don't taste the cream by accident — it looks and smells just too inviting! Allow the mixture to rest and cool down, until it starts to become solid. Now, whip it up — ideally, with a hand-whisk — until the body butter is fluffy and light, just like chocolate mousse. The perfect after-bath dessert for your skin. Apply plentifully as a daily treat.

SUPERPOWERS EXPLAINED

Cocoa butter improves skin elasticity and accelerates the process of collagen production. Shea butter is anti-inflammatory, moisturising and supports the skin's own healing process. It is beneficial in treating eczema, scars and skin blemishes. Coconut oil has amazing moisturising and nourishing properties. Cocoa powder is rich in antioxidants, while its antiseptic powers will clean and heal skin blemishes. Cinnamon has antibacterial properties. The invigorating eucalyptus essential oil generally purifies the body and gives this cream a delightfully fresh edge.

SHELF LIFE
6 MONTHS
STORE IN A COOL, DARK PLACE

CLEANSING FACIAL OIL

Cleansing oil may sound strange at first, but once this ancient remedy has found its way onto your face, you will not want anything else for your evening beauty routine. Coconut oil acts as a nourishing base and combined with castor oil, which acts as a purifier and cleanser, it will leave you with a healthy glow.

– KARIN

THINGS YOU'LL NEED

4 tablespoons coconut oil

1 tablespoon castor oil

1 teaspoon rosehip oil

20 drops of lavender essential oil

10 drops of chamomile blue essential oil or chamomile Roman essential oil

GET STARTED

Mix the coconut, castor and rosehip oils together, and stir well for a few minutes. Now add the essential oils, drop by drop, mixing carefully. Transfer to a glass jar with a lid for daily use.

Pour a small amount of the oil mixture into your hands and gently massage onto your face in circular motions for 4–5 minutes. Hold a fine cotton washcloth under very hot running water, then wring it out quickly and place the steaming washcloth over your face for about 1 minute to let the steam open the pores and cleanse your skin of impurities. Gently remove the remaining oil from your face with the washcloth - there's no need to use soap. Your skin will feel soft and nourished.

SUPERPOWERS EXPLAINED

Coconut oil has amazing moisturising and nourishing properties with antibacterial and anti-inflammatory qualities. Castor oil purifies, cleanses and encourages collagen production while penetrating deep into the skin. Rosehip oil is very high in antioxidants and helps regenerate damaged skin tissue. Lavender and chamomile blue essential oils have anti-inflammatory, antibacterial, antiviral and skin-soothing powers.

SHELF LIFE
4–6 MONTHS
STORE IN A COOL, DARK PLACE

EMPRESS SISI'S SHINY HAIR RINSE FOR GIRL BOSSES

This old Austrian remedy was used religiously by Empress Elisabeth – 'Sisi' – of Austria, who was well-known for her beauty and charisma. Besides being one of the first women famous for a gym-like exercise regime, a fierce horse rider and being endlessly busy running a country, Sisi was also obsessed with her hair, which is impressive with a to-do list as packed as hers. This is one of her many hair-care routines for creating volume and shine. To be used whenever you need to channel the fierce girl boss you are.

– NICI

THINGS YOU'LL NEED

2 egg yolks

2½ tablespoons Cognac

1½ tablespoons raw apple cider vinegar

GET STARTED

Whisk the egg yolks and Cognac together until smooth.

Work the mixture into your clean, wet hair, brushing it through with a wide-toothed comb.

Rinse out with lukewarm water. Then rinse your hair out a second time, but this time add the apple cider vinegar to the water running over your hair (for extra shine). And that's it: an Empress' beauty routine!

SUPERPOWERS EXPLAINED

Egg yolks add protein and strength to fine hair. Cognac adds brightness and shine to hair and creates lots of volume. Apple cider vinegar detangles and de-frizzes hair, balances pH levels and gives hair shine.

SHELF LIFE
USE RIGHT AWAY
ENOUGH FOR ONE APPLICATION

DIY DEODORANT

'Body odour tackled in a non-toxic way' is our motto for this recipe. Having prepared some of the mixtures in this book, you will already be familiar with most of the ingredients and getting started will not make you sweat!

– KARIN

THINGS YOU'LL NEED

2 tablespoons beeswax

2 tablespoons shea butter

2 tablespoons coconut oil

30 drops of lavender essential oil

8 drops of cypress essential oil

8 drops of lemon essential oil

8 drops of tea tree essential oil

8 drops of sage essential oil (do not use sage essential oil during pregnancy)

GET STARTED

Gently heat the beeswax in a double boiler until liquid. Heat the shea butter in a separate pan, then add the coconut oil, stirring gently and being careful not to overheat the mixture. Finally, add the liquid beeswax.

As soon as the ingredients have blended, remove from the heat and stir continuously until the mixture starts to become milky. Once the mixture has cooled down a little, add the essential oils while continuing to stir, and then very quickly pour into a convenient container for easy application and leave to cool.

SUPERPOWERS EXPLAINED

Beeswax doesn't block pores and has antibacterial and deeply moisturising benefits. Shea butter has moisturising, anti-inflammatory and wound-healing properties. Coconut oil is nourishing, anti-microbial, antifungal, antibacterial and anti-inflammatory. Lavender, tea tree and sage essential oils are antibacterial, antiviral and antifungal, the latter also being an antiperspirant. Lemon and cypress essential oils are cleansing and astringent, and cypress essential oil is also perspiration-regulating.

SHELF LIFE
6-8 MONTHS
STORE IN A COOL, DARK PLACE

BLEMISH-CURING SUNFLOWER SEED MASK

Our skin is our biggest organ and takes in all the good and the bad we apply to it. Therefore, I like the idea of only feeding my skin with ingredients that are safe enough to be eaten! This face mask follows this rule. It will help the sensitive skin on your face to clear away impurities.

– NICI

THINGS YOU'LL NEED

1 teaspoon honey

1 teaspoon olive oil

60 g (2 oz/½ cup) sunflower seeds, ground

GET STARTED

Warm up the honey in a double boiler. Once it is smooth and runny, take off the heat and add the olive oil. Stir gently until they are both completely blended together.

Now stir in the ground sunflower seeds. Let the mixture cool so it is not too hot for your face. Apply a thick layer onto clean skin, relax and allow the mask to work its miracles. And no nibbling on the face mask! After 20–30 minutes, wash off the mask and moisturise your skin with one of our nourishing face creams.

SUPERPOWERS EXPLAINED

Honey has wound-healing, anti-microbial and antibacterial properties. Olive oil is antibacterial and anti-inflammatory, protects and nourishes skin, and is high in vitamins, antioxidants and minerals. Sunflower seeds are full of a whole host of vitamins, including vitamins A, B and E, as well as the beneficial fatty acids.

SHELF LIFE
USE RIGHT AWAY
ENOUGH FOR ONE APPLICATION

ANTI-DANDRUFF BIRCH TONIC

Give your hair a birching - but in a good way! Use the leaves to make this invigorating infusion to soothe your scalp and alleviate the symptoms of dandruff.

- KARIN

THINGS YOU'LL NEED

250 ml (8½ fl oz/1 cup) raw apple cider vinegar

2 tablespoons birch leaves

30 drops of cedar wood essential oil

GET STARTED

Bring the vinegar to the boil in a pot covered with a lid.

Put the birch leaves in a cup and pour the hot vinegar over the leaves. Cover, then leave to infuse for 15 minutes.

Strain the brew and let it cool before mixing in the cedar wood essential oil.

After washing your hair, massage 3–4 tablespoons of tonic gently into your scalp. No need to rinse.

Store the remaining tonic in a glass bottle.

SUPERPOWERS EXPLAINED

Raw apple cider vinegar is antibacterial, antifungal, clarifying, cleansing and a natural conditioner. It also restores the pH balance of the scalp, which is important because frequent use of unhealthy products or foods can lead to an imbalance in our bodies. Birch leaves encourage blood circulation and soothe itching. Cedar wood essential oil is wound- and scar-healing, regenerating, analgesic, antibacterial, antifungal and anti-inflammatory. Both birch leaves and cedar wood essential oil have anti-dandruff qualities.

SHELF LIFE
1 MONTH
KEEP REFRIGERATED

ENERGY-BOOSTING COFFEE BODY SCRUB

Coffee's noble ability to wake me up makes it my favourite commodity. With this recipe we make its uplifting skill-set go even further – in combination with my next ever-so-favourite flavour: chocolate. Those two culinary superheroes result in a rich and deliciously fragrant scrub that will leave your skin just yummy.

– NICI

THINGS YOU'LL NEED

2 tablespoons cocoa butter

2 tablespoons coconut oil

2 tablespoons jojoba oil

6 tablespoons ground coffee

2 teaspoons cocoa powder

20 drops of peppermint essential oil

GET STARTED

Slowly melt the cocoa butter and coconut oil in a double boiler. Once melted, take off the heat and stir in the jojoba oil.

Mix in the coffee grounds and cocoa powder. Lastly, stir in the essential oil; peppermint will give this delicacy for your skin a refreshing edge.

The scrub will be even more invigorating if kept cool in the fridge! Rub the scrub all over your body up to once a week. Always rinse the scrub off well afterwards. We love to use this scrub in the morning or before going out, as it is uplifting and boosts our energy.

SUPERPOWERS EXPLAINED

Cocoa butter improves elasticity of the skin and boosts collagen production. Coconut oil has amazing moisturising and nourishing properties. Jojoba oil will lock moisture into your skin. Coffee, besides the lovely smell, improves circulation and blood flow, and its small particles are a perfect exfoliant to remove dry skin. Cocoa powder is high in antioxidants, whilst its antiseptic powers will clean as well as heal skin blemishes. Peppermint essential oil is not only anti-inflammatory and cooling, it also stimulates blood flow and is detoxifying.

SHELF LIFE
2 WEEKS
KEEP REFRIGERATED

NOURISHING
RASPBERRY
FACE MASK

If you are not already eating fresh raspberries regularly, please start now. They are so good for you! Once you have been convinced by the outstanding benefits of fresh raspberries, it won't take much persuasion to squash them and apply them to your face. Easy and quick to prepare, this mask will leave your skin glowing and pleading for more!

– KARIN

THINGS YOU'LL NEED

10 fresh raspberries

GET STARTED

Pick your raspberries fresh in the summer or remove 10 raspberries from the fridge 15–30 minutes before use.

Squash the raspberries and apply immediately to your skin. Leave for 15 minutes, then wash off gently with tepid water.

SUPERPOWERS EXPLAINED

Raspberries contain powerful antioxidants and have anti-inflammatory properties.

SHELF LIFE
USE RIGHT AWAY
ENOUGH FOR ONE APPLICATION

HARD-WORKING HAND LOTION

If you are cursed with granny-like hands, this is for you. I think I have another 30 years until my actual age will catch up with the look of my hands. (Maybe I'm exaggerating the tiniest bit.) Fair enough though: a lot of horse riding, home renovating, mountain climbing and being generally hands-on hasn't helped to maintain a youthful look for my hands — and I don't mind being reminded of all those adventures. So, let's not forget to reward our hands for all their hard work — and we love to use this nourishing recipe to do so.

– NICI

THINGS YOU'LL NEED

½ tablespoon beeswax

1 tablespoon cocoa butter

1 tablespoon shea butter

1 tablespoon coconut oil

½ teaspoon vegetable glycerin

2 teaspoons purified warm water

12 drops of neroli essential oil

8 drops of jasmine absolute oil

GET STARTED

Gently heat the beeswax in a double boiler. Add the cocoa butter, being careful not to overheat the mixture. Then add the shea butter, followed by the coconut oil and glycerin. Stir in the warm water until it is all mixed together, then remove from the heat.

While the mixture is cooling down, stir continuously until it becomes milky. Blend with a stick blender until the oils and water have emulsified. Gently stir in the drops of beautifully smelling oils and pour into a glass jar. Let the cream cool down before using it to pamper your hands as often as needed.

SUPERPOWERS EXPLAINED

Cocoa butter improves elasticity of the skin and accelerates the process of collagen production. Shea butter's wound-healing, anti-inflammatory and moisturising qualities make it perfect for this cream. It is also beneficial in treating scars and blemishes. Coconut oil is moisturising and nourishing and neroli essential oil nourishes dry skin. Jasmine absolute oil not only smells amazing, but is also anti-spasmodic, analgesic, nourishing and healing. Vegetable glycerin is hydrating; beeswax is healing and moisturising.

SHELF LIFE
4–6 MONTHS
STORE IN A COOL, DARK PLACE

NETTLE HAIR TONIC FOR STRENGTH & BALANCE

This tonic won't sting; it will soothe and leave your hair and scalp nourished and healthy instead. Nettles and raw apple cider vinegar combined: what a wonderful way to get back to nature.

– KARIN

THINGS YOU'LL NEED

4 teaspoons nettle leaves

500 ml (17 fl oz/2 cups) boiling water

2 tablespoons raw apple cider vinegar

GET STARTED

Pour the boiling water over the nettle leaves, cover and leave to infuse for 15 minutes.

Strain the liquid, leave to cool down and then add the vinegar. Rinse your hair several times with the mixture after shampooing, while massaging the tonic into your scalp gently. There's no need to rinse afterwards.

SUPERPOWERS EXPLAINED

Nettles contain lots of vitamins and minerals, which nourish and strengthen the hair roots, promote circulation, alleviate an itchy scalp and stimulate hair growth. They are also high in antioxidants. Raw apple cider vinegar is antibacterial, antifungal, clarifying, cleansing and a natural conditioner. It also restores the pH balance of the scalp, which is important because frequent use of unhealthy products or foods can lead to an imbalance.

SHELF LIFE
USE RIGHT AWAY
ENOUGH FOR ONE RINSE

'HOLIDAY GLOW' SEA SALT BODY SCRUB

There is nothing like a day at the beach to give your skin sun-kissed softness from a good dive into the salty sea. When you crave that feeling without a Caribbean beach just around the corner, this scrub will do the trick and give your skin a holiday-like glow, minus the tan, but hey, it's the 21st century and tans are not so hot these days. (Seriously, protect your skin: wear sunscreen!)

– NICI

THINGS YOU'LL NEED

1 tablespoon coconut oil

2 tablespoons avocado oil

3 tablespoons sea salt

10 drops of peppermint essential oil

5 drops of eucalyptus essential oil

3 drops of lemon essential oil

3 drops of sage essential oil (do not use sage essential oil during pregnancy)

GET STARTED

Melt the coconut oil in a double boiler. Once it has softened, take off the heat and stir in the avocado oil. Let the mixture cool down.

Add the sea salt and carefully drop in the essential oils. Stir well. Transfer to a glass jar and keep the scrub in a cool place. Use this scrub once a week all over your body for smooth and glowing skin. Always rinse the scrub off well after the treatment.

SUPERPOWERS EXPLAINED

Coconut oil has amazing moisturising and nourishing properties. Avocado oil is full of protein and fats, both great for strong and moisturised skin and hair. Ideal for dry skin, it is also rich in vitamins C, E and K as well as magnesium and potassium. Sea salt is a disinfectant and full of minerals. Peppermint essential oil improves focus and boosts energy. Eucalyptus essential oil is invigorating and generally purifies the body. Sage essential oil is stimulating, invigorating, antiperspirant, antibacterial, antiviral and antifungal, wound-healing and skin-regenerating. Lemon essential oil is cleansing and high in vitamin C.

SHELF LIFE
4–6 MONTHS
STORE IN A COOL, DARK PLACE

HEAVENLY FACE CREAM

Shea butter and jojoba oil have been cherished since ancient times for their skin-healing and deeply moisturising properties. Cold-pressed oils from these superfruits form the base of our face cream. We've then mixed them with a few more exceptional ingredients so it smells to-die-for! It will be hard to believe you produced this heavenly concoction yourself.

– KARIN

THINGS YOU'LL NEED

1 tablespoon beeswax

1 tablespoon shea butter

2½ tablespoons jojoba oil

2 teaspoons rosehip oil

10 drops of frankincense essential oil

4 drops of immortelle essential oil

2 drops of jasmine absolute oil

GET STARTED

Carefully and slowly melt the beeswax in a double boiler on a low heat until liquid.

Heat the shea butter in a separate bowl, also using a double boiler. Then add the jojoba and rosehip oils, stirring gently and being careful not to overheat the mixture.

Once the beeswax has melted and both liquids have a similar temperature, add the oil mix to the beeswax and stir well. As soon as the ingredients are blended, remove from the heat and stir continuously until the cream has cooled down a little and starts to become milky.

At this point, add the essential oils and and jasmine absolute oil and stir well. Pour into small pots.

SUPERPOWERS EXPLAINED

Beeswax is a natural emulsifier and preservative and is healing and moisturising. Shea butter is anti-inflammatory, deeply moisturising and wound-healing. Jojoba oil unclogs pores, is anti-inflammatory, antibacterial, wound-healing and rich in vitamin E. Rosehip oil is high in anti-oxidants, very hydrating and healing. Frankincense and immortelle essential oils promote the growth of healthy new skin cells, are wound-healing, anti-inflammatory, antibacterial and antiviral. Jasmine absolute oil is nourishing and healing.

SHELF LIFE
4–6 MONTHS

DESERT-DRY HAIR MIRACLE MASK

The idea of putting avocado in your hair may sound a bit extreme, but really you should have thought about that before bleaching your hair white to dye it pastel pink, then bleaching it again to dip-dye it mint green, followed by a visit to the hairdresser where you demanded, 'Something that looks natural'. By saying you, I mean me, obviously. But it's nothing that can't be fixed by a bit of protein-rich hair TLC.

– NICI

THINGS YOU'LL NEED

½ avocado

3 tablespoons coconut oil

2 tablespoons avocado oil

15 drops of rosemary essential oil

GET STARTED

Peel the avocado half, mash it and set aside.

Melt the coconut oil in a double boiler. Once melted, take off the heat and stir in the avocado oil.

Next, mix in the mashed avocado. Once everything is smooth, carefully add the essential oil, drop by drop.

No need to wash your hair beforehand - simply apply the hair mask straight to your hair, cover with a shower cap and wrap in a pre-heated towel. Now ... wait! Best use the waiting time to dream up some future hair colours. Keep the mask on for about 20 minutes, then carefully comb through sections of your hair with a wide-toothed comb. Once your mane feels a bit more tamed, rinse thoroughly. Try not to blow-dry your hair, as this will dry it out again. We recommend a beachy air-dried look: find your inner wild child.

SUPERPOWERS EXPLAINED

Avocado is full of proteins, vitamins C, E and K, and fats, which nurture your hair. Cconut oil has moisturising and nourishing properties. Rosemary essential oil regulates sebum for the scalp and encourages blood flow.

SHELF LIFE
USE RIGHT AWAY
ENOUGH FOR ONE APPLICATION

ALL NATURAL BEAUTY

SOOTHING POTATO SLICES FOR TIRED, PUFFY EYES

We love them baked or steamed, as a side or as a main. But who would have thought that potatoes can be used raw and are the star of the show in their own right? Try this for puffy eyes — the only question: slivered or sliced?

– KARIN

THINGS YOU'LL NEED

2 × 5 mm (¼ in) thick potato slices

GET STARTED

Place a potato slice on each eye. Rest for 15 minutes and then rinse with warm water. Apply once or twice a day.

SUPERPOWERS EXPLAINED

Potatoes are anti-inflammatory, detoxifying, antiseptic and have soothing and decongesting properties. They are high in potassium and vitamins B and C and also contain calcium and iron.

SHELF LIFE
USE RIGHT AWAY
ENOUGH FOR ONE APPLICATION

SOLID WAX PERFUME

Do we love this most for its gorgeous smell or the fact that the perfume is solid? You decide! The waxy consistency allows this perfume to be fitted into the tiniest container and snugly hidden in the smallest of purses for a top-up any time.

– NICI

THINGS YOU'LL NEED

1 tablespoon beeswax

1 tablespoon coconut oil

7 drops of rose absolute oil

7 drops of cedar wood essential oil

7 drops of bergamot essential oil

5 drops of jasmine absolute oil

2 drops of black pepper essential oil

GET STARTED

Melt the beeswax in a double boiler. When it has nearly melted, add the coconut oil. Stir gently until both have turned into a smooth liquid.

Remove from the heat and carefully stir in the essential oils, one by one. Before the perfume sets, pour into the containers of your choice. Simply rub a bit of the perfume onto your skin to enjoy the luscious smell.

SUPERPOWERS EXPLAINED

Beeswax is moisturising and coconut oil has amazing moisturising and nourishing properties – it also has a delicious smell, which contributes to this perfume's scent. The base note, rose absolute oil, adds a beautiful floral accent to this perfume. Jasmine absolute oil has a refreshing and relaxing effect on the mind and emotions, and is a highly valued ingredient in perfumes. Cedar wood essential oil is a base note with an uplifting effect on your mood. The top note, bergamot essential oil, is both relaxing and uplifting, and often used in eau de cologne. The middle-note, black pepper essential oil, is surprisingly fresh, spicy and woody, and has energising properties.

SHELF LIFE
1 YEAR
STORE IN A COOL, DARK PLACE

REPAIRING MASK FOR DAMAGED HAIR

One thing you won't be able to get enough of once you reap the benefits for lifeless hair: egg yolk! This hair mask is made from the most unusual blend of ingredients. Unbelievably nourishing, this goo will render your hair radiant and revitalised.

– KARIN

THINGS YOU'LL NEED

3-4 tablespoons castor oil

2 egg yolks

45 g (1½ oz/¼ cup) fresh yeast

juice of 1 lemon

GET STARTED

Mix the caster oil with the egg yolks. Add the yeast and lemon juice, and stir well to form a smooth paste.

Massage gently and evenly through dry, unwashed hair. Leave for 20–30 minutes, well-covered with a towel or a shower cap, or both.

Rinse out with tepid water and wash your hair with a mild shampoo.

SUPERPOWERS EXPLAINED

Castor oil purifies, cleanses and encourages collagen production while penetrating deep into the skin. Egg yolk is high in proteins, vitamins and minerals. Yeast contains many minerals, trace elements and vitamins. Lemon is cleansing, high in vitamin C and flavonoids and has antibacterial, antiviral and anti-inflammatory properties. It also encourages collagen production while toning the scalp.

SHELF LIFE · ENOUGH FOR ONE APPLICATION
USE RIGHT AWAY

SOOTHING FOOT BATH FOR ACHING FEET

It's summer, it's boiling and for some unjust reason you aren't at the beach, which clearly must mean a bigger plan has gone horribly wrong. Time for a refreshing foot bath, which also works wonderfully for feet that have been tortured on a high-heeled night out. This foot bath is so refreshing, as well as wound-healing for those times your shoes have left your feet red and throbbing.

– NICI

THINGS YOU'LL NEED

3 litres (10 fl oz/12 cups) water

handful of fresh thyme leaves

handful of fresh peppermint leaves

small piece of ginger (approx 1 in/2.5 cm) peeled and chopped

GET STARTED

Bring the water to the boil in a large pan. Add the herbs and ginger and simmer for a few minutes. Take off the heat and leave to infuse for 2 hours.

Strain the liquid into a footbath or foot-sized bowl, then soak your feet for a blissful 20 minutes, while sitting down in a comfy chair (this is important). After the time is up, gently dry your feet and apply a light cream (such as the one on page 80). Now, feet up!

SUPERPOWERS EXPLAINED

Thyme has soothing and wound-healing properties. Peppermint is not only antibacterial and antiviral, but also anti-inflammatory, strongly analgesic and cooling. It also stimulates the blood flow and is detoxifying. Ginger stimulates blood circulation.

SHELF LIFE
USE RIGHT AWAY
ENOUGH FOR ONE APPLICATION

REPLENISHING COCONUT OIL SCALP BALM

The healing powers of coconut oil have been underestimated for far too long. This remedy treats dandruff and relieves an itchy scalp. Make it a 2-in-1 elixir by combing this highly valued mix of oils through your hair to replenish lost elasticity.

– KARIN

THINGS YOU'LL NEED

3 tablespoons coconut oil

8 drops of rosemary essential oil

8 drops of cedar wood essential oil

GET STARTED

Mix together the coconut oil and the essential oils, stir well and store the mixture in a glass jar.

Gently massage the balm into a dry scalp and wrap your head in a towel or shower cap.

For even more impact, you can also comb the balm through your hair to nourish and moisturise.

Relax for 30–60 minutes or leave overnight before washing your hair and scalp with a mild shampoo.

SUPERPOWERS EXPLAINED

Coconut oil is highly moisturising, nourishing and has anti-microbial, antifungal, antibacterial and anti-inflammatory properties. Rosemary essential oil is antibacterial and antiviral, regulates sebum for your scalp and encourages blood flow. Cedar wood essential oil is wound- and scar-healing, regenerating, analgesic, antibacterial, antifungal and anti-inflammatory.

SHELF LIFE
1 YEAR
STORE IN A COOL, DARK PLACE

'BOTTLED HOLIDAY' BODY CREAM

So we took an orange orchard, some Mediterranean sunshine and then added some silky softness. And, voilà – here it is: the perfect body cream, which carries the smell of a holiday in a bottle.

– NICI

THINGS YOU'LL NEED

½ tablespoon beeswax

1 tablespoon shea butter

1 tablespoon coconut oil

2 tablespoons jojoba oil

12 drops of bergamot essential oil

7 drops of neroli essential oil

7 drops of jasmine absolute oil

GET STARTED

Start by melting the beeswax and shea butter in a double boiler. Once they are nearly melted, stir in the coconut oil. Take off the heat once the wax, butter and oil have melted into a smooth liquid.

Now, carefully stir in the jojoba oil. Add all the essential oils last, whilst the mix is cooling down. Mix well until the cream is smooth.

We hope that a little bit of a holiday feel will be released whenever you use this body cream, and the smell will transport you right to a sunshine-filled orange grove.

SUPERPOWERS EXPLAINED

Beeswax is skin-protecting and moisturising. Shea butter is wound-healing and moisturising and supports the skin's own healing process. Coconut oil has deeply moisturising and nourishing properties. Jojoba oil is great for locking in moisture – it even offers natural sun protection, but only up to factor 4 (so you will still need to use your regular sunscreen). Besides the beautiful smells, the essential oils contain important properties: bergamot supports healthy skin, neroli nourishes dry skin, and jasmine absolute oil is nourishing and healing.

SHELF LIFE
4–6 MONTHS
STORE IN A COOL, DARK PLACE

REPAIRING HAIR OIL

Argan oil was one of the best-kept secrets of North Africa, and only recently has this mysterious oil found its way to other parts of the world. Its superb qualities will nourish and improve the texture of your hair. It may not make it as strong as Rapunzel's, but this oil will do wonders for damaged and colour-treated locks.

– KARIN

THINGS YOU'LL NEED

2 tablespoons argan oil

3 drops of rosemary essential oil

3 drops of cedar wood essential oil

3 drops of lavender essential oil

GET STARTED

Pour the argan oil into a small bottle and add the essential oils. Shake well and use a tiny amount on the tips of wet hair after washing to bring moisture to dry ends.

Alternatively, you can use it as an intense conditioner to bring back shine. Massage the mixture into tired, dry and lifeless hair. Leave it to absorb for 15–20 minutes before jumping into the shower and shampooing.

SUPERPOWERS EXPLAINED

Argan oil is highly moisturising, nourishing and rich in antioxidants and vitamins A and E. It stimulates cell activity and boosts circulation. Rosemary essential oil is antibacterial and antiviral, sebum-regulating for your scalp and encourages blood flow. Cedar wood essential oil is wound- and scar-healing, regenerating, analgesic, antibacterial, antifungal and anti-inflammatory. Lavender essential oil has anti-inflammatory, antibacterial, antiviral, antifungal, analgesic and skin-soothing powers and helps to heal wounds and scars.

SHELF LIFE
1 YEAR
STORE IN A COOL, DARK PLACE

ALL NATURAL BEAUTY

REPAIRING CUTICLE BUTTER

Busy hands that are used to craft, make, mend and create, or that simply have been cursed by the evil demon called Dry Skin, might need a bit of extra loving cuticle-care to look as neat and healthy as you want them. Worry not: here is a bit of repairing magic for those with stubbornly dry skin.

– NICI

THINGS YOU'LL NEED

2 tablespoons shea butter

1 tablespoon coconut oil

2 teaspoons honey

2 teaspoons argan oil

20 drops of rosehip oil

10 drops of rose absolute oil

GET STARTED

Heat the shea butter and coconut oil in a double boiler. Stir in the honey whilst it all melts together. Remove from the heat and stir in the argan oil.

Now add the essential oils, drop by drop. Wait until the cream slowly starts to solidify as it cools, then it's time to whip it all up — best to do this with a hand-whisk. Whisk away, until the cream thickens to a light and spreadable consistency. There you go, all done — the ultimate rescue for your cuticles whenever they need some extra love!

SUPERPOWERS EXPLAINED

Shea butter is anti-inflammatory and supports the skin's own healing process, which makes it perfect for this cream. The nourishing coconut oil has anti-inflammatory properties. Honey is wound-healing and antibacterial, while argan oil is rich in antioxidants and vitamins A and E. Rosehip oil helps to regenerate damaged skin tissue. Rose absolute oil is known to have rejuvenating properties.

SHELF LIFE
4–6 MONTHS
STORE IN A COOL, DARK PLACE

NOURISHING GREEN TEA TONER

The green tea we love is cultivated in a sunny spot in Japan and harvested around May. The most delicate and tender leaves are gently steamed or roasted and then rolled, breaking the cells and releasing flavours unique to one of the healthiest teas in the world. While you're preparing a refreshing drink, this toner allows you to spoil your skin at the same time.

– KARIN

THINGS YOU'LL NEED

1 teaspoon organic green tea or Japanese Sencha Uchiyama green tea

250 ml (8½ fl oz/1 cup) water

optional: a few drops (¼ teaspoon) raw apple cider vinegar

GET STARTED

Boil the kettle, then let the water cool down to 80°C (175°F). Be precise; we are trying to get the most out of this tea.

Pour the water over the leaves into a cup, cover and leave to infuse for 8–10 minutes.

Strain the leaves and allow the brew to cool, to avoid burning your face. Add the raw apple cider vinegar, if using, and stir well.

Use straight away as a toner for your face before applying moisturiser, and drink the rest. We drink a lot of green tea and this has become a daily routine.

SUPERPOWERS EXPLAINED

Green tea in general is very good for you, but Sencha Uchiyama green tea is a special kind. It is very high in antioxidants, is anti-inflammatory, antifungal, antibacterial and draws out impurities. Brew it according to the above instructions and the infusion will award you the most polyphenols and catechins of all green teas. Raw apple cider vinegar is antibacterial, antifungal, clarifying and cleansing. It also restores the pH balance of the skin, which is important because frequent use of unhealthy products or foods can lead to an imbalance.

SHELF LIFE
USE RIGHT AWAY
ENOUGH FOR ONE APPLICATION

'SHINY MANE' ROSEMARY HAIR RINSE

For a shiny, healthy-looking mane, the answer lies in your herb patch. No matter if you plant rosemary and chamomile in the garden of your epic stately home, or in a little herb pot balanced on the smallest of windowsills, those fragrant herbs are worth growing just for this remedy.

– NICI

THINGS YOU'LL NEED

1 litre (34 fl oz/4 cups) water

5 tablespoons rosemary, fresh or dried

5 tablespoons chamomile, fresh or dried

GET STARTED

Bring the water to the boil in a saucepan and add the rosemary and chamomile. Take off the heat and allow to infuse for 2 hours.

Strain the liquid and rinse your hair with the infusion at the end of your regular hair-care routine. Shine!

SUPERPOWERS EXPLAINED

Rosemary stimulates circulation in the scalp and increases shine, while chamomile helps to relieve irritated and dry scalps. Both have antibacterial properties.

SHELF LIFE
USE RIGHT AWAY
ENOUGH FOR ONE APPLICATION

WHIPPED BODY CREAM

You might see the world through rose-coloured glasses after makiing this cream. It is so simple to make, requires no heating, is packed with antioxidants, locks in your skin's moisture and you will end up smelling of roses. What's not to like?

– KARIN

THINGS YOU'LL NEED

50 ml (2 fl oz) jojoba oil

100 g (3½ oz) shea butter

15 drops of rose absolute oil

35 drops of cedar wood essential oil

GET STARTED

Mix the jojoba oil and shea butter together in a glass jar and whisk with a fork for a while. Surprisingly quickly, the mixture will start to become a heavenly whipped cream. Keep mixing until all the cream is completely smooth.

Now add the rose absolute oil and cedar wood essential oil and stir well.

SUPERPOWERS EXPLAINED

Jojoba oil is absorbed very easily into the skin and provides some natural sun protection, but only up to factor 4 (so you will still need to use your regular sunscreen). It is wound-healing, anti-inflammatory, antibacterial and rich in antioxidants and vitamin E. In addition, it treats sunburns, scars and eczema. Shea butter is wound-healing, anti-inflammatory, moisturising and supports the skin's own healing process. It is beneficial in treating eczema, scars and skin blemishes. Cedar wood essential oil is wound- and scar-healing, antibacterial, antifungal and anti-inflammatory. Rose absolute oil is antibacterial, antiviral and is known to have cell-regenerating and wound-healing properties.

SHELF LIFE
1 YEAR
STORE IN A COOL, DARK PLACE

BEACH WALK
FOOT SCRUB

Time to spoil your old hoofs let's pretend they are on holiday. Get your neglected feet out of their woolly socks and give them the beachy TLC they so deserve.

– NICI

THINGS YOU'LL NEED

3 tablespoons honey

3 tablespoons sea salt

1 tablespoon coconut oil

GET STARTED

Mix all the ingredients together, then apply to your feet and scrub! Make sure to sit down for this part, as it is a ticklish affair and a slippery matter. Best to rinse both feet and your bathtub thoroughly before making a move, to avoid a spectacular 'I-slipped-in-my-bath' selfie moment!

SUPERPOWERS EXPLAINED

Honey has wound-healing, anti-microbial and antibacterial properties. Sea salt is disinfecting, full of minerals and, of course, makes an ideal exfoliant. Coconut oil has amazing moisturising and nourishing properties, and is also anti-microbial, antifungal, antibacterial and anti-inflammatory.

SHELF LIFE
1 YEAR
STORE IN A COOL, DARK PLACE

NOURISHING CHICKPEA & TURMERIC FACE MASK

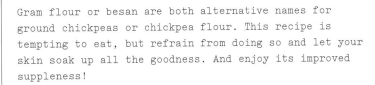

Gram flour or besan are both alternative names for ground chickpeas or chickpea flour. This recipe is tempting to eat, but refrain from doing so and let your skin soak up all the goodness. And enjoy its improved suppleness!

– KARIN

THINGS YOU'LL NEED

1 tablespoon chickpea flour

2 teaspoons almond oil

2 teaspoons honey

2 teaspoons lemon juice

1 teaspoon ground turmeric

GET STARTED

Mix together all the ingredients and stir well to form a paste. Apply a thick layer onto clean skin and rest for 15 minutes before rinsing off the mask with tepid water.

Don't worry about turmeric colouring your face; the mask washes off very easily without a trace.

SUPERPOWERS EXPLAINED

Chickpeas and turmeric both have antioxidant properties. Furthermore, turmeric is known to be antibacterial, anti-inflammatory and wound-healing. Almond oil calms irritation, relieves inflammation, is deeply moisturising and soothing. Honey has wound-healing and antibacterial properties. Lemon is cleansing, has antibacterial and anti-inflammatory properties and also encourages collagen production while toning the skin.

SHELF LIFE
7 DAY
KEEP REFRIGERATED

'EASY REPAIR' HOT OIL HAIR TREATMENT

THINGS YOU'LL NEED

125 ml (4 fl oz/½ cup) coconut oil

125 ml (4 fl oz/½ cup) avocado oil

My seemingly indestructible quarter-Italian (thanks to Grandpa Paul) mane has a lot to deal with: bleach and dye, salty sea (on lucky days) and never enough time to use conditioner. Time for some repair work - quick and simple.

- NICI

GET STARTED

Heat the coconut oil in a double boiler. When the coconut oil is nearly melted, stir in the avocado oil. Once it is a smooth liquid, take it off the heat.

To avoid burns, carefully check the temperature of the mixture before applying. The mask should be warm but comfortable on your skin — never too hot.

Work the hot oil mixture through your dry hair, covering it from roots to tips. Then put on a shower cap and wrap a heated towel around your head — or find a bit of sunshine to sit in — to keep the hair mask warm for at least half an hour. Final step: wash your hair with a gentle shampoo.

SUPERPOWERS EXPLAINED

Coconut oil has amazing moisturising and nourishing properties. Avocado oil is full of proteins and fats, both great to nurture and strengthen your hair. Ideal for dry skin and hair, it is rich in vitamins C, E and K as well as magnesium and potassium.

SHELF LIFE
USE RIGHT AWAY
ENOUGH FOR ONE APPLICATION

RELAXING BATH OIL

It's the end of the day and you deserve a rest. Fill your bathtub with sea salt, inhale the essences and let your thoughts take you far, far away.

– KARIN

THINGS YOU'LL NEED

2 tablespoons almond oil

a handful of sea salt

5 drops of carrot seed essential oil

5 drops of myrrh essential oil

2–3 drops of rose otto essential oil

GET STARTED

Mix the almond oil with the sea salt and essential oils.

Add the salty, oily mixture to running bath water before stepping in. Give the water a good whirl and enjoy for 30 minutes.

SUPERPOWERS EXPLAINED

Almond oil contains many fatty acids, minerals and vitamins. It calms irritation, is anti-inflammatory, deeply moisturising and soothing. Sea salt is disinfecting and full of minerals. Rose otto essential oil is antibacterial, antiviral, antispasmodic, cell-regenerating and has wound-healing properties. Carrot seed essential oil has anti-inflammatory, skin-regenerating and wound-healing properties. It also lowers blood pressure and aids the healing of burns. Myrrh essential oil is antibacterial, antiviral, anti-inflammatory, skin-regenerating, nourishing, hormone-regulating and wound-healing.

SHELF LIFE
USE RIGHT AWAY
ENOUGH FOR ONE APPLICATION

MOISTURISING AVOCADO FACE MASK

Winter is coming and it's time for comfort foods — foods to comfort your dry winter skin. A simple trip to the fruit bowl and this deeply moisturising face mask is as good as ready to rescue you from an itchy, flaky, central-heating-induced salamander mid-peel look.

– NICI

THINGS YOU'LL NEED

¼ avocado

¼ banana

1 teaspoon honey

1 teaspoon avocado oil

GET STARTED

Mash the avocado and the banana. Mix with the rest of the ingredients until completely smooth — this will be easiest with a hand-mixer (or just use a whisk and a strong forearm).

Generously apply to your cleansed face and neck, relax and feel your skin replenish with goodness. After 20 minutes, wash off with lukewarm water.

SUPERPOWERS EXPLAINED

Avocado oil and raw avocado are full of protein and fats. Ideal for dry skin, they are rich in vitamins C, E and K, as well as magnesium and potassium. Honey has wound-healing, anti-microbial and antibacterial properties. Banana is rich in moisturising potassium and vitamin E to combat free radicals.

SHELF LIFE
USE RIGHT AWAY
ENOUGH FOR ONE APPLICATION

SOFTENING LEMON FOOT BATH

Apart from lifting your mood, this unbelievably soothing and softening foot bath will make the skin on your feet feel really happy. You won't know how you dealt with calluses before. Happiness from head to toe!

– KARIN

THINGS YOU'LL NEED

juice of 2 lemons

1 teaspoon olive oil

GET STARTED

Get a foot bath of warm water ready, then add the lemon juice and stir well before soaking your feet.

Bathe your feet for 20–30 minutes. Any hard skin on your feet will come off very easily when you start scrubbing. As a finishing touch, moisturise your feet with the olive oil. Your feet will become softer and smoother if you use this foot bath regularly.

SUPERPOWERS EXPLAINED

Lemon is cleansing, has antibacterial, antiviral and anti-inflammatory properties and encourages collagen production. Olive oil is antibacterial and anti-inflammatory, protects and nourishes skin, and is high in vitamins, antioxidants and minerals.

SHELF LIFE
USE RIGHT AWAY
ENOUGH FOR ONE SESSION

'WAKE ME UP' BODY LOTION

Some say that grapefruit can minimise cellulite, but we don't quite trust such claims as cellulite is caused too deep beneath the skin to be reached by any cream. So forget that cellulite even exists: just own your dimples and celebrate your lovely curves with this beautiful cream! The invigorating smell of grapefruit and peppermint will wake you up after a morning shower and help to get you going. More often than not, this uplifting smell even wakes me up enough to go for a morning run — now *that* will help to fight cellulite — maybe we should have named this cream 'anti-cellulite' after all!

– NICI

THINGS YOU'LL NEED

4 tablespoons shea butter

4 tablespoons rosehip oil

25 drops of grapefruit essential oil

25 drops of cypress essential oil

GET STARTED

Heat the shea butter in a double boiler, stirring while it melts. Take it off the heat and stir in the rosehip oil. Carefully stir in the essential oil.

We recommend that you use this refreshing and uplifting body lotion whenever you feel like you need a bit of an energy boost.

SUPERPOWERS EXPLAINED

Shea butter is moisturising and supports the skin's own healing process. The rosehip oil in this cream is extremely high in antioxidants and penetrates deep into the skin to stimulate collagen production. The refreshing grapefruit essential oil supports metabolism. Cypress essential oil improves circulation and reduces varicose veins.

SHELF LIFE
USE RIGHT AWAY
ENOUGH FOR ONE APPLICATION

DIVINE MOISTURISING FACIAL OIL

Originally from Morocco, argan oil has superb nutritional qualities for skin and hair. Combined with tamanu oil, also known as the oil of the gods, this facial oil will make your skin feel divine.

– KARIN

THINGS YOU'LL NEED

4 teaspoons argan oil (or jojoba oil)

2 teaspoons tamanu oil (or rosehip oil)

2 drops of jasmine absolute oil

6 drops of chamomile Roman essential oil

GET STARTED

Mix the argan and tamanu oils together in a small glass bottle. Add the jasmine absolute oil and chamomile essential oil and shake gently.

Apply a few drops to your face in slow circular motions as part of your daily routine. Avoid contact with eyes.

SUPERPOWERS EXPLAINED

Argan oil is highly moisturising and nourishing and has a rejuvenating effect. It stimulates cell activity, boosts circulation and is rich in antioxidants and vitamins A and E. Tamanu oil has wonderful wound-healing qualities, promotes the formation of new skin tissue and has anti-inflammatory, antioxidant and antibacterial properties. It is also high in vitamin E. Jasmine absolute oil is soothing and has a strong mood-lifting effect. Chamomile Roman essential oil is anti-inflammatory, skin-regenerating and wound-healing. Optional jojoba oil and rosehip oil are high in antioxidants, anti-inflammatory, moisturising and wound-healing.

SHELF LIFE
4–6 MONTHS
STORE IN A COOL, DARK PLACE

WHIPPED FIG
BODY SCRUB

Figs are not just delicious with some goat's cheese, they are also an amazingly fragrant superfruit, ideal to use in beautifying skincare rituals. The figs' tiny seeds are the gentle exfoliators in this rich and nourishing scrub. It will leave your skin so smooth and soft you won't be able to resist – you'll strike a pose and demand: 'Draw me like one of your French girls.'

– NICI

THINGS YOU'LL NEED

2 large ripe figs

1 tablespoon shea butter

1 tablespoon beeswax

1 tablespoon coconut oil

4 tablespoons brown sugar

GET STARTED

Peel and mash the figs, then set aside.

Start to melt the shea butter and beeswax over a double boiler. When they are nearly melted, stir in the coconut oil. Once melted together, take off the heat, stir in the mashed figs and mix well.

Let the mixture cool down a bit before adding the sugar (otherwise the sugar will melt). Best to use a hand-whisk to add the sugar to give the scrub a lovely light and feathery finish.

Use this gentle, exfoliating body scrub once a week for deliciously soft skin you'll want to show off.

SUPERPOWERS EXPLAINED

Shea butter is moisturising and supports the skin's own healing process – a great base for this scrub. The skin-protecting and moisturising properties of the beeswax are as important as the nourishing effects of the coconut oil. Most importantly for this scrub, figs have a hydrating effect on the skin and are known to be antioxidant and anti-inflammatory. The figs' small seeds together with the sugar grains work superbly as an exfoliant.

BALANCING KAOLIN POWDER MIX

Used regularly, kaolin powder helps maintain a healthy balance of oils on the scalp and skin. Kaolin powder gently cleanses by attaching itself to oily and dirty compounds and is easily rinsed off, helping to reduce the need for conventional soaps and shampoos. If you have dry and sensitive skin or damaged hair this might be just what you have been looking for!

– KARIN

THINGS YOU'LL NEED

5 tablespoons kaolin powder

150 ml (5 fl oz/1½ cups) warm water

GET STARTED

Mix the powder with the warm water to form a thick, smooth paste.

Massage the paste into your scalp and hair in the same way as you would with a conventional shampoo. Leave for a few minutes to absorb. Rinse with warm water. After regular use, the need to wash your hair will become less frequent.

Used as a body wash it will leave your skin soft and cleansed while not drawing out the skin's own oils.

KAOLIN FACE MASK

Used as a face mask it purifies, nourishes and boosts skin cell renewal. For each 1 tablespoon of kaolin, mix with 1 tablespoon of green tea to make a paste. Apply the paste to your face, rest for 10-15 minutes, then wash off with warm water.

SUPERPOWERS EXPLAINED

Kaolin powder, when mixed with water, absorbs oily particles and dirt. It is full of minerals and nutrients that protect and nourish the skin. It stimulates circulation and is antibacterial and de-acidifying. Green tea is high in polyphenols, and has anti-inflammatory and antibacterial properties.

SHELF LIFE
USE RIGHT AWAY
ENOUGH FOR ONE APPLICATION

HERBAL WARDROBE FRESHENER

Scent can be such a powerful thing; it instantly transports us back to forgotten times, places and people. It can even affect our mood. Or, in more humble times, a nice scent just strong enough to overpower your gym kit's smell in your wardrobe, is exactly the luxury you need.

– NICI

THINGS YOU'LL NEED

bunch of fresh rosemary

bunch of fresh sage

bunch of fresh eucalyptus

small white cotton bag

20 drops of basil essential oil

2 cotton balls

20 drops of lavender essential oil

GET STARTED

Roughly chop the fresh herbs so that they fit inside the cotton bag.

Place the drops of basil oil on one of the cotton balls, and the lavender oil on the other cotton ball.

Place the herbs together with the cotton balls in the cotton bag. Hang the bag inside your wardrobe and enjoy the meadow-fresh scent taking over!

SUPERPOWERS EXPLAINED

Rosemary has a strong aromatic and Mediterranean smell. Sage has been a popular ingredient throughout the history of perfume due to its leathery scent. Eucalyptus is invigorating and generally purifying. The sweet and spicy basil essential oil helps relieve the symptoms of stress. Lavender essential oil also has a long history in perfume-making, with a floral, fresh and strong smell.

SHELF LIFE
1 MONTH
REPLACE MONTHLY

MOISTURISING & HEALING LIP BALM

With shea butter as its base — considered a superior healer — this lip balm will turn chapped and tender lips into lustrous ones before you know it. Rosehip oil, with its amazing skin-rejuvenating properties, is the ideal addition to this exceptional moisturiser.

– KARIN

THINGS YOU'LL NEED

1 tablespoon beeswax

4 teaspoons shea butter

1 tablespoon argan oil

1 teaspoon rosehip oil

3 drops of melissa essential oil

3 drops of mandarin essential oil

3 drops of myrrh essential oil

GET STARTED

Gently melt the beeswax in a double boiler on a low heat until liquid.

In a separate bowl, also using a double boiler, heat the shea butter, then add the argan and rosehip oils, stirring gently and being careful not to overheat the mixture. Next, add the beeswax. As soon as the ingredients have blended together, remove from the heat and stir continuously to cool down the mixture.

When it starts to become milky, add the essential oils while continuing to stir. Quickly pour the balm into small pots before the beeswax hardens.

SUPERPOWERS EXPLAINED

Beeswax has skin-protecting and moisturising properties. Shea butter is wound-healing, anti-inflammatory and moisturising. Argan oil is highly moisturising, nourishing and has a rejuvenating effect, stimulates cell activity, is rich in antioxidants and vitamins A and E. Rosehip oil is moisturising, extremely high in anti-oxidants, has anti-inflammatory properties and helps to heal wounds and soften scars. Melissa essential oil is anti-inflammatory and antiviral. Mandarin essential oil is mood-elevating and strengthens the immune system. Myrrh essential oil is antibacterial, antiviral, anti-inflammatory, nourishing and wound-healing, especially around the mouth and throat.

SHELF LIFE
4–6 MONTHS
STORE IN A COOL, DARK PLACE

'HANGOVER-FIGHTING' BODY OIL

So, you have a hangover and it's not pretty. We recommend after the first obvious healing steps of tea and bacon sarnie, ideally delivered to your bed, and the soothing chanting of 'I will never drink again', it's time for step two: hop in the shower to wash away the hangover. Step three is to treat yourself to this body oil that will leave you feeling pampered, refreshed and maybe even so uplifted that you will be brave enough to face your phone to see whom you drunk-texted.

– NICI

THINGS YOU'LL NEED

3 tablespoons jojoba oil

3 tablespoons almond oil

8 drops of eucalyptus essential oil

8 drops of peppermint essential oil

8 drops of lavender essential oil

4 drops of tea tree essential oil

4 drops of geranium essential oil

GET STARTED

This is the easiest of all recipes: simply combine all the oils in a glass bottle and shake well until they are mixed and ready.

Use this body oil after a refreshing shower and, voilà, hangover, no more!

SUPERPOWERS EXPLAINED

Eucalyptus essential oil is invigorating and generally purifies the body. Peppermint essential oil supports digestion, improves focus, boosts energy, reduces fever, clears headaches and is used for muscle pain relief. Lavender essential oil has anti-inflammatory, anti-spasmodic, antibacterial and antiviral properties and is soothing for the skin. Tea tree essential oil reduces bad odours and can help stimulate the immune system. Geranium essential oil helps skin stay healthy and is known to revitalise body tissues. It heals inflamed skin and other minor skin damage. Rich in antioxidants and vitamin E, jojoba oil is anti-inflammatory, antibacterial and has a disinfecting effect. Almond oil contains fatty acids, minerals and vitamins. It calms irritation, is anti-inflammatory, deeply moisturising and soothing – and soothing is what you need, when battling a hangover.

SHELF LIFE
4–6 MONTHS
STORE IN A COOL, DARK PLACE

ROYAL REJUVENATING OATMEAL FACE MASK

Nearly 150 years ago, Austrian Empress Sisi's beauty regime included getting up at 6 am to take a cold bath, followed by a full body massage, then routines of creams for body, face and hair, after which came jogging and gymnastics. Sisi was known to put more food on her body than she ate.

This iconic face mask recipe is very simple. To start with it's similar to porridge, then just add rosewater or rose essential oil and you are there. Breakfast and beauty in one: it doesn't get any easier than this.

– KARIN

THINGS YOU'LL NEED

1 tablespoon oats or oatmeal

2 tablespoons warm milk (or water or soya milk)

2 teaspoons rose water or 2 drops of rose absolute oil or rose otto essential oil

GET STARTED

Mix the oats or oatmeal with the warm milk. Stir for 4–5 minutes to form a thick paste before slowly adding the rose water or essential oil.

Apply evenly to your face and leave for up to 30 minutes before rinsing off with tepid water.

SUPERPOWERS EXPLAINED

Oats have anti-inflammatory and antioxidant properties, promote wound-healing and have a calming and healing effect on skin, while milk is moisturising, nourishing and soothing. Rose water, rose absolute oil and rose otto essential oil are antibacterial, antiviral and have cell-regenerating and wound-healing properties.

SHELF LIFE
USE RIGHT AWAY
ENOUGH FOR ONE APPLICATION

DETOXIFYING MATCHA TEA FACE MASK

No, it's not dessert. And this wonderful green paste will only make you look like the Wicked Witch of the West temporarily. Once your skin has absorbed the nutrients you will feel refreshed and beautiful! Enjoy!

-KARIN

THINGS YOU'LL NEED

1 teaspoon matcha tea powder

1 teaspoon yoghurt

GET STARTED

Put the matcha tea powder in a bowl and stir in the yoghurt to make a smooth paste.

Apply to a cleansed face and relax for 15 minutes. Remove gently with tepid water and enjoy a healthy glow.

SUPERPOWERS EXPLAINED

Matcha tea is high in vitamins, antioxidants, polyphenols, flavonoids and minerals. It is anti-inflammatory, antibacterial and draws out impurities. Yoghurt is moisturising and nourishing.

SHELF LIFE
USE RIGHT AWAY
ENOUGH FOR ONE APPLICATION

KEY
INGREDIENTS

A little collection of the ingredients in this book and their superpowers explained.

Almond oil: contains fatty acids, minerals and vitamins. It calms irritation, is anti-inflammatory, deeply moisturising and soothing.

Aloe vera: anti-inflammatory, antibacterial, antiviral, moisturising and cell-regenerating.

Apple cider vinegar: antibacterial, antifungal, clarifying, cleansing and a natural conditioner. It also restores the pH balance of the skin.

Argan oil: highly moisturising, nourishing and rich in antioxidants and vitamins A and E. It stimulates cell activity and boosts circulation.

Avocado oil: contains proteins and fats, avocado oil is ideal for dry skin. It is rich in vitamins C, E and K, magnesium and potassium.

Banana: rich in moisturising potassium and vitamin E to combat free radicals.

Beeswax: natural preservative and emulsifier, healing, moisturising, skin protecting.

Bergamot essential oil: relaxing, antibacterial, antiviral, antispasmodic.

Birch leaves: encourage blood circulation and soothe itching, and have anti-dandruff qualities.

Black pepper essential oil: antibacterial, anti-inflammatory, has energizing properties.

Carrot seed essential oil: anti-inflammatory, skin-regenerating and wound-healing.

Castor oil: purifies, cleanses, improves skin elasticity and encourages collagen production.

Cedar wood essential oil: wound- and scar-healing, regenerating, analgesic, antibacterial, antifungal.

Chamomile: disinfects and has anti-inflammatory, antibacterial and antispasmodic properties. It also helps to relieve irritated and dry scalps.

Chamomile blue essential oil: antibacterial, anti-inflammatory, anti-spasmodic and anti-fungal with skin-soothing and wound-healing powers.

Chamomile Roman essential oil: anti-inflammatory, anti-spasmodic, wound-healing, skin regenerating.

Cocoa butter: improves skin elasticity and accelerates the process of collagen production.

Cocoa powder: antiseptic, rich in antioxidants and cleans and heals skin blemishes.

Coconut oil: has amazing moisturising and nourishing properties with antibacterial and anti-inflammatory qualities.

Coffee: improves circulation and blood flow and is a perfect exfoliant.

Cypress essential oil: purging, anti-spasmodic, astringent, improves circulation and reduces varicose veins.

Egg yolk: high in proteins, vitamins and minerals.

Eucalyptus essential oil: anti-bacterial, anti-viral, antifungal, invigorates and purifies the body.

Figs: have a hydrating effect on the skin and are known to be antioxidant and anti-inflammatory.

Frankincense essential oil: wound-healing, anti-spasmodic, anti-inflammatory, antibacterial, antiviral, stimulates blood flow and promotes skin-cell growth.

Geranium essential oil: anti-inflammatory, antibacterial, astringent, anti-fungal, wound-healing, helps skin stay healthy and is known to revitalise body tissues.

Ginger: anti-inflammatory, antiviral, disinfecting, skin regenerating, stimulates the immune system and blood circulation.

Grapefruit essential oil: astringent, anti-spasmodic, disinfecting, refreshing and stimulates metabolism.

Green tea: high in polyphenols and catechins, and has anti-inflammatory and antibacterial properties.

Honey: strengthens the immune system, prevents bacterial growth and reduces fever. Also wound-healing, nutritious and invigorating but sensitive to heat. Use the most natural honey you can find: organic, raw honey is best.

Immortelle essential oil: disinfecting, wound-healing, anti-inflammatory, antibacterial, antiviral, stimulates blood flow and promotes the growth of new skin cells.

Jasmine absolute oil: anti-spasmodic, analgesic, nourishing, soothing and mood lifting.

Jojoba oil: unclogs pores, is anti-inflammatory, antibacterial, wound-healing, easily absorbed into the deepest layers of the skin, hydrophilic and rich in vitamin E.

Kaolin powder: absorbs oily particles and dirt when mixed with water. It is full of minerals and nutrients that protect and nourish the skin. It stimulates circulation, is antibacterial and is de-acidifying.

Lavender: anti-inflammatory, anti-bacterial, antiviral, anti-spasmodic, antifungal, pain-relieving, wound-healing, relaxing and soothing.

Lemons: antibacterial, antiviral, body-cleansing, infection-fighting, rich in vitamin C and flavonoids. The juice also acts as a preservative.

Mandarin essential oil: mood elevating, strengthens the immune system and anti-spasmodic.

Matcha tea: high in vitamins, antioxidants, polyphenols, flavonoids and minerals. It is anti-inflammatory, antibacterial and draws out impurities.

Melissa essential oil: anti-inflammatory, antiviral, strengthens the immune system, is anti-spasmodic and pain relieving.

Myrrh essential oil: antibacterial, antiviral, astringent anti-inflammatory, skin-regenerating, nourishing, hormone-regulating and wound-healing.

Neroli essential oil: antibacterial, antiviral, antifungal, anti-spasmodic, wound-healing, alleviates itching and nourishes dry skin.

Nettles: act as a tonic, are anti-allergenic, nourishing, strengthening, anti-spasmodic, promote circulation, alleviate itching, stimulate hair-growth, are blood-cleansing, blood-building, high in vitamin C and iron.

Oak bark: antiperspirant, antibacterial, antiviral, anti-inflammatory, astringent and alleviates itching.

Oats: have anti-inflammatory and antioxidant properties, promote wound-healing and have a calming and healing effect on skin.

Olive oil: anti-bacterial, anti-inflammatory, protects and nourishes skin, contains antioxidants, minerals and vitamins.

Peppermint essential oil: anti-inflammatory, anti-bacterial, anti-viral, wound-healing, alleviates an itch and is strongly cooling, sebum regulating, stimulates the blood flow and is detoxifying.

Potatoes: anti-inflammatory, detoxifying, antiseptic and have soothing and decongesting properties. They are high in potassium and vitamins B and C and also contain calcium and iron.

Raspberries: contain powerful antioxidants and have anti-inflammatory properties.

Rose absolute oil: antibacterial, antiviral, antispasmodic, soothing and mood elevating. It is known to have rejuvenating properties and promotes a glowing complexion.

Rose otto essential oil: antibacterial, antiviral, antispasmodic, strengthens the immune system, is pain-relieving, soothing, relaxing, cell-regenerating, rejuvenating and has wound-healing properties.

Rosehip oil: very high in important fatty acids and helps to regenerate damaged skin tissue. It stimulates collagen production, is sebum regulating, moisturising, anti-inflammatory and wound-healing.

Rosemary essential oil: anti-bacterial, antiviral, sebum regulating, stimulates blood flow and metabolism.

Sage: antibacterial, fungicidal, anti-viral, wound-healing, antiperspirant and skin regenerating.

Sea salt: disinfecting, full of minerals and makes an ideal exfoliant.

Shea butter: anti-inflammatory, hygroscopic, deeply moisturising, wound-healing and supports the skin's own healing process. It is beneficial in treating eczema, scars and skin blemishes.

Sunflower seeds: full of a whole host of vitamins, including vitamins A, B and E, as well as beneficial fatty acids.

Tamanu oil: is wound-healing, promotes the formation of new skin tissue and has anti-inflammatory, antioxidant and antibacterial properties. It is also high in vitamin E.

Tea tree essential oil: antibacterial, antifungal, antiviral, anti-inflammatory, skin-regenerating, antiperspirant, pain relieving, stimulates the blood-flow, wound-healing, alleviates itching and strengthens the immune system.

Thyme: antibacterial, fungicidal, anti-viral, wound-healing, pain relieving, stimulates the blood-flow and strengthens the immune system.

Turmeric: antibacterial, anti-inflammatory and wound-healing.

Vegetable glycerin: hydrating and emulsifying.

Yeast: contains many minerals, trace elements and vitamins.

GLOSSARY

Absolute oil: oily mixture extracted from plants using solvent extraction techniques.

Analgesic: pain-relieving.

Anti-microbial: destructive to or inhibiting the growth of micro-organisms.

Antioxidants: Substances that have the potential to prevent oxidative damage to cell membranes, DNA and other cellular macromolecules, when free radical activity outweighs the cell's own antioxidant defence mechanisms.

Antispasmodic: loosens muscle spasms.

Astringent: causing contraction of soft tissue.

Catechins: specific flavonoids which are antifungal, antibacterial and anti-carcinogenic.

Collagen: fibrous protein found in skin, bones, muscles, tendons and other connective tissue.

Emulsifier: an agent that binds immiscible liquids.

Essential oil: oily mixture extracted from plants using steam distillation extraction techniques.

Flavonoids: a sub group of phytonutrients or polyphenols, some of which are anti-allergenic, anti-inflammatory, antioxidant, anti-microbial.

Fungicidal: destructive to the growth of fungi.

Polyphenols: secondary metabolites or phytonutrients found in large quantities in plants. Some are anti-allergenic, anti-inflammatory, anti-fungal, antibacterial, antioxidant, anti-microbial and have anti-carcinogenic properties.

Sebum: an oily secretion of the sebaceous glands.

BEAUTY TREATMENT INDEX

Page references in *italics* are recipe entries

INDEX

Page references in *italics* are
recipe entries

THANK YOUS
DANKESCHÖNS

Our special thanks go to our family and friends who have supplied us with many recipes, tirelessly supported, tested and critiqued our products and potions to make them what they are in this book. We are eternally grateful to Tony for his continued support for our project — all the recipes were shot in his studio.

Our special special thanks go to the wonderful team at Hardie Grant. To Stephen, Kate and Kajal who supported our project from the start with excitement. To Laura for the many changes and suggestions. Finally, to Nicky, who managed to put the book together so beautifully. We are thrilled with how it looks — thank you so much.

ABOUT
THE
AUTHORS

Karin and Nici had to live apart in Austria before they finally met in London, where they have worked together on an array of projects in advertising, photography and illustration ever since.